ISBN 978-1-333-23553-6
PIBN 10477275

This book is a reproduction of an important historical work. Forgotten Books uses
state-of-the-art technology to digitally reconstruct the work, preserving the original format
whilst repairing imperfections present in the aged copy. In rare cases, an imperfection in
the original, such as a blemish or missing page, may be replicated in our edition. We do,
however, repair the vast majority of imperfections successfully; any imperfections that
remain are intentionally left to preserve the state of such historical works.

1 MONTH OF
FREE
READING

at
www.ForgottenBooks.com

By purchasing this book you are eligible for one month membership to ForgottenBooks.com, giving you unlimited access to our entire collection of over 700,000 titles via our web site and mobile apps.

To claim your free month visit:
www.forgottenbooks.com/free477275

English
Français
Deutsche
Italiano
Español
Português

www.forgottenbooks.com

Mythology Photography **Fiction**
Fishing Christianity **Art** Cooking
Essays Buddhism Freemasonry
Medicine **Biology** Music **Ancient**
Egypt Evolution Carpentry Physics
Dance Geology **Mathematics** Fitness
Shakespeare **Folklore** Yoga Marketing
Confidence Immortality Biographies
Poetry **Psychology** Witchcraft
Electronics Chemistry History **Law**
Accounting **Philosophy** Anthropology
Alchemy Drama Quantum Mechanics
Atheism Sexual Health **Ancient History**
Entrepreneurship Languages Sport
Paleontology Needlework Islam
Metaphysics Investment Archaeology
Parenting Statistics Criminology
Motivational

A Decision Procedure for Set-Theoretic
Formulae Involving Rank
and Cardinality Comparison

D. Cantone
V. Cutello

Technical Report 440

April 1989

A decision procedure for set-theoretic formulae involving rank and cardinality comparison *

D. CANTONE, V. CUTELLO
Department of Computer Science
Courant Institute of Mathematical Sciences
New York University
251 Mercer Street, New York, N.Y. 10012, U.S.A., and
Department of Mathematics
University of Catania
Viale Andrea Doria, 6A, 95125 Catania, Italy

1 Introduction

In this paper we show that the satisfiability problem for a sublanguage of set theory involving the notion of rank and cardinality comparison is decidable. We recall that given a set, its rank is a measure of the nesting level of elements within it, whereas its cardinality is the number of its elements (cf. [Jec78]).

The present result combines and extends decision methods developed in [CCF88] and [CC88] with integer linear programs (see for example [GN72] and [Sal75] for various integer linear programs algorithms).

Our motivation in developing decision tests for sublanguages of set theory lies in a long term project for the design and implementation of a set-theoretically based proof verifier. Decision procedures of the kind described in $\mathcal{T}$heorem 3.1 (see below) should constitute the inferential core of such a system (see [CS88]).

The universe $\mathcal{V}$ of sets we will consider is the standard von Neumann universe, satisfying, among others, the *axiom of foundation:*

"*every non-empty set has an element which is disjoint from it.*"

One of the consequences of this axiom is that there cannot exist sets $s_0, s_1, \ldots, s_n$, with n any natural number, such that

$$s_0 \in s_1 \in \cdots \in s_n \in s_0 .$$

It also follows that our universe $\mathcal{V}$ is stratified according to the following recursive definition

$$\mathcal{V}_0 = \emptyset$$

*This work has been partially supported by ENI and ENIDATA within the AXL project.

$$\mathcal{V}_\alpha \;=\; \bigcup_{\beta<\alpha} pow(\mathcal{V}_\beta)$$

$$\mathcal{V} \;=\; \bigcup_{\alpha\in Ord} pow(\mathcal{V}_\alpha)\,,$$

where Ord is the class of all ordinal numbers. Then the rank of a set s, denoted by $rank(s)$, can be defined as the minimum ordinal α such that $s \subseteq \mathcal{V}_\alpha$ (see [Jec78] for further details).

The class of formulae we will consider in this paper is denoted MLSSRC, an acronym for *Multi-Level Syllogistic with Singleton, Rank comparison, and Cardinality comparison.* MLS is the class of unquantified set-theoretic formulae in the language $=, \in, \cup, \cap, \setminus$ together with propositional connectives; its satisfiability problem has been solved in [FOS80]. The intended meaning of the singleton operator $\{\cdot\}$ and the predicates rank comparison $\leq$, $<$, and cardinality comparison $|\ |_\leq$, $|\ |_<$ is the following

- $x = \{y\}$ is true if y is the only member of x;

- $x \leq y$ [resp. $x < y$] is true if and only if $rank(x) \leq rank(y)$ [resp. $rank(x) < rank(y)$];

- $x|\ |_{\leq y}$ [resp. $x|\ |_{<y}$] is true if and only if $|x| \leq |y|$ [resp. $|x| < |y|$], i.e. if and only if the number of elements in x is less than or equal to [resp. less than] the number of elements in y. For simplicity we will write $|x| \leq |y|$ [resp. $|x| < |y|$] in place of $x|\ |_{\leq y}$ [resp. $x|\ |_{<y}$].

By a simple normalization procedure (cf. [Can88], [CC88]) it can easily be shown that the satisfiability problem for the class of formulae MLSSRC is equivalent to the *injective* satisfiability problem for conjunctions of literals each of which has one of the following types:

$$\begin{array}{ll}
(=) & x = y \cup z \,,\; x = y \setminus z \\
(\{\cdot\}) & x = \{y\} \\
(\leq, <) & x \leq y \,,\; x < y \\
(|\ |_\leq, |\ |_<) & |x| \leq |y| \,,\; |x| < |y| \,,
\end{array}$$

where we recall that a formula is injectively satisfiable if it has a 1-1 model. We will call such formulae *normalized conjunctions* of MLSSRC.

The decidability of the extension of MLS with the singleton operator, the cardinality operator together with arithmetic addition, subtraction and comparison was proved in [FOS80]. Also, the decidability of MLS extended with the singleton operator and the rank comparison predicate was established in [CCF88]. This paper extends and unifies both such results.

Let P be any normalized conjunction of MLSSRC. The following notion of *place of P* is of central importance in what follows (cf. [CFMS87], [CS88], [CCF88]; see also [Can88] and [CC88] for an extensive bibliography).

Definition 1.1 *Given a conjunction P, a place of P is any 0/1-valued function π defined on the set of variables of P and such that $\pi(x) = \pi(y) \vee \pi(z)$ [resp. $\pi(x) = \pi(y) \wedge \neg\pi(z)$] whenever $x = y \cup z$ [resp. $x = y \setminus z$] occurs in P. (Here we are obviously identifying 0 and 1 with the truth values false and true, respectively.)* ∎

2 Preliminary definitions

In order to give the decision test for MLSSRC, we need a bit of terminology. Let P be a normalized conjunction of MLSSRC. Let $V = \{y_1, y_2, \ldots, y_m\}$ be the collection of distinct variables occuring in P. Let also $\Pi = \{\pi_1, \ldots, \pi_n\}$ be a set of places for P.

Remark 2.1 In the following we will freely identify variables y_s and places π_i with their indices s and i, respectively. ∎

Definition 2.1 *We say that a place π_i is a* singleton place *if there exist y_s, y_t such that $y_s = \{y_t\}$ is in P and $\pi_i(y_s) = 1$.*
   We denote by $SING$ the set of singleton places.

To each variable y_s we associate the set of places $\Pi(y_s)$ defined by

$$\Pi(y_s) = \{i : \pi_i(y_s) = 1\}. \tag{1}$$

Moreover, given a map $F : \{1, 2, \ldots, m\} \to \{1, 2, \ldots, n\}$, to each place π_i we associate the set of variables $V_F(\pi_i)$ defined by

$$V_F(\pi_i) = \{s : F(s) = i\} \tag{2}$$

(in the following, when the map F is fixed, we will simply write V in place of V_F).

Definition 2.2 *An* admissible set of filled places *(with respect to a set of places Π and a map F) is any subset $\mathcal{F}$ of Π such that:*

(a) $SING \subseteq \mathcal{F}$; and

(b) if $i \in \mathcal{F}$ then $V(\pi_i) \neq \emptyset$.

   A place π_i is said to be filled *(with respect to $\mathcal{F}$) if $i \in \mathcal{F}$.*

Given an increasing sequence of integers $r_0 = 0 < r_1 < \ldots < r_\ell = n$, we define a map $R : \{1, \ldots, n\} \to \{1, \ldots, \ell\}$ by putting

$$R(i) = \min\{h : r_{h-1} < i \leq r_h\}. \tag{3}$$

Clearly R is nondecreasing.
   We also define a map * on V by putting

$$s^* = \max\{R(j) : j \in \Pi(y_s)\}. \tag{4}$$

Definition 2.3 *An* admissible set of trapped places *(with respect to a set of places Π, a map F, an admissible set of filled places $\mathcal{F}$, and an increasing sequence of integers $r_0 = 0 < r_1 < \ldots < r_\ell = n$) is any subset T of Π such that:*

(a) if $i \in T$, then $i' \in T$, for all $i' \in \{1, \ldots, r_{R(i)}\}$; and

(b) if $i \in \mathcal{F}$ and $\Pi(y_s) \subseteq T$ for all $s \in V(\pi_i)$, then $i \in T$.

A place π_i is said to be trapped *(with respect to T) if $i \in T$.*
A variable y_s is said to be trapped *(with respect to T) if $\Pi(y_s) \subseteq T$.*

Definition 2.4 *An admissible set of finite places (with respect to a set of places Π, a map F, an admissible set of filled places $\mathcal{F}$, an increasing sequence of integers $r_0 = 0 < r_1 < \ldots < r_\ell = n$, and an admissible set of trapped places T) is any subset FIN of Π such that*

(a) $\mathcal{F} \cup T \subseteq FIN$;

(b) if $\Pi(y_t) \subseteq FIN$ and either $|y_s| \leq |y_t|$ or $|y_s| < |y_t|$ is in P, then $\Pi(y_s) \subseteq FIN$.

A place π_i is said to be finite *(with respect to FIN) if $i \in FIN$.*

3   The main result

We are now ready to state our main theorem.

Theorem 3.1 *Let P be a normalized conjunction of MLSSRC, whose distinct variables are $V = \{y_1, \ldots, y_m\}$. Then P is injectively satisfiable if and only if there exist*

- *a set $\Pi = \{\pi_1, \ldots, \pi_n\}$ of places of P,*
- *a map $F : \{1, \ldots, m\} \rightarrow \{1, \ldots, n\}$,*
- *an admissible set of filled places $\mathcal{F}$,*
- *an increasing sequence of integers $0 = r_0 < r_1 < \ldots < r_\ell = n$,*
- *an admissible set of trapped places $T = \{1, \ldots, r_{k_1}\} \subseteq \Pi$ and an assignment of sets $\overline{\pi}$ of rank less than or equal to k_1 to the trapped places π,*
- *an admissible set of finite places $FIN \subseteq \Pi$, and*
- *a map $C : INF \rightarrow \{0, 1, \ldots, n - f - 1\}$, where $INF = \Pi \setminus FIN$ and $f = |FIN|$,*

such that the following conditions are satisfied:

Condition (C1). *No two distinct variables of P are Π-equivalent.*

Condition (C2). *The partial assignment defined over trapped variables y by*

$$\overline{M}y = \bigcup_{\pi(y)=1} \overline{\pi}$$

is an injective model for the literals of P involving only trapped variables. Moreover, if y_i and π_j are trapped and $\overline{M}y_i \in \overline{\pi}_j$, then $F(i) = j$.

Condition (C3). *If $\pi_j(y_s) = 1$, then $R(j) < R(F(s))$ (where R is defined as in (3)).*

Condition (C4). *If $y_s = \{y_t\}$ is in P, then $\Pi(y_s) = \{F(t)\}$.*

Condition (C5).

(C5.1) If $y_s \leq y_t$ [resp. $y_s < y_t$] is in P, then $s^* \leq t^*$ [resp. $s^* < t^*$] (where * is the map defined in (4)).

(C5.2) If $i \in \mathcal{F}$, then $R(i) = \max\{s^* : s \in V(\pi_i)\} + 1$.

Condition (C6). If $|y_s| \leq |y_t|$ [resp. $|y_s| < |y_t|$] is in P and $\Pi(y_s) \cap INF \neq \emptyset$, then

$$\max\{\mathcal{C}(j) : j \in \Pi(y_s) \cap INF\} \leq \max\{\mathcal{C}(j) : j \in \Pi(y_t) \cap INF\}$$

$$[\text{resp. } \max\{\mathcal{C}(j) : j \in \Pi(y_s) \cap INF\} \leq \max\{\mathcal{C}(j) : j \in \Pi(y_t) \cap INF\}].$$

(Notice that if $\Pi(y_s) \cap INF \neq \emptyset$, then $\Pi(y_t) \cap INF \neq \emptyset$.)

Condition (C7). Let ξ_i, $i \in FIN$, be distinct integer variables. Then the following system SYS of equations and inequalities in the unknowns ξ_i has a positive integer solution:

(C7.1) if $i \in FIN \setminus \mathcal{F}$ [resp. $i \in \mathcal{F}$], then

$$\xi_i > |V(\pi_i)|$$

$$[\text{resp. } \xi_i = |V(\pi_i)|]$$

is in SYS;

(C7.2) if $|y_s| \leq |y_t|$ [resp. $|y_s| < |y_t|$] is in P and $\Pi(y_t) \subseteq FIN$ then

$$\sum_{i \in \Pi(y_s)} \xi_i \leq \sum_{j \in \Pi(y_t)} \xi_j$$

$$\left[\text{resp. } \sum_{i \in \Pi(y_s)} \xi_i < \sum_{j \in \Pi(y_t)} \xi_j\right]$$

is in SYS;

(C7.3) for each $i \in T$ the equation

$$\xi_i = |\bar{\pi}_i|$$

is in SYS.

In the next subsections we will show that conditions (C1)-(C7) are necessary and sufficient for P to be injectively satisfiable.

Theorem 3.1 contains a decision test for MLSSRC, since all its conditions are algorithmically verifiable. In fact, places of P, admissible sets of filled places, etc., vary over a finite family of objects depending solely on the conjunction P and a priori determinable. In addition, given any admissible set of trapped places $T = \{1, \ldots, r_{k_1}\} \subseteq \Pi$, there can be only finitely many different assignments of sets $\bar{\pi}$ of rank less than or equal to k_1 to the trapped places π, since $\mathcal{V}_{k_1}$ is finite. Finally, condition (C7) can be tested by using any algorithm for solving integer linear programs (see, for example, [Sal75], [GN72]). Therefore we have the following decidability result:

Corollary 3.1 *The class MLSSRC has a solvable satisfiability problem.*

3.1 Necessity of conditions (C1)-(C7)

We begin by showing that conditions (C1)-(C7) of Theorem 3.1 are necessary for a normalized conjunction P of MLSSRC to be injectively satisfiable. So, let P be such a conjunction and let M be a 1-1 model for P. Let also $V = \{y_1, \ldots, y_m\}$ be the set of variables occurring in P. Consider the disjoint regions $\sigma_1, \ldots, \sigma_n$ of the Venn diagram of the sets $My_1, \ldots, My_m$ in the universe

$$\mathcal{U} = My_1 \cup \ldots \cup My_m \cup \{My_1, \ldots, My_m\}.$$

Clearly each region σ determines a place π of P defined on V by putting

$$\pi(x) = \begin{cases} 1 & \text{if } \sigma \subseteq Mx \\ 0 & \text{if } \sigma \cap Mx = \emptyset. \end{cases}$$

Let Π be the set of all such places. Notice that $My_s = \bigcup_{\pi_j(y_s)=1} \sigma_j$, for each variable y_s in P. Therefore, if x, y are distinct variables of P, by the injectivity of M, $Mx \neq My$, so that there exists a place $\pi \in \Pi$ such that

$$\pi(x) = 1 \text{ if and only if } \pi(y) = 0.$$

Hence condition (C1) is satisfied.

Next, for each variable y_s, let σ^{y_s} denote the region of the Venn diagram which contains My_s as an element. Then we define the map $F : \{1, \ldots, m\} \rightarrow \{1, \ldots, n\}$ by putting

$$F(s) = j \quad \text{if and only if} \quad My_s \in \sigma_j \ (\equiv \sigma^{y_s}). \tag{5}$$

We also put

$$\mathcal{F} = \{j : |\sigma_j| = |V(\sigma_j)|\}.$$

Clearly, $\mathcal{F}$ is an admissible set of filled places.

Without loss of generality, we can assume that the sets $\sigma_1, \ldots, \sigma_n$ are indexed in such a way that if $rank(\sigma_i) < rank(\sigma_j)$, then $i < j$. This determines uniquely an increasing sequence of integers $r_0 = 0 < r_1 < \ldots < r_\ell = n$ such that for all $i, i' \in \{1, \ldots, n\}$

$$rank(\sigma_i) = rank(\sigma_{i'}) \text{ iff } r_{h-1} < i, i' \leq r_h, \text{ for some } h \in \{1, \ldots, \ell\}.$$

To find an admissible set of trapped places, we use the following procedure.

> **Proc Find_Trap;**
> $\quad T \leftarrow \emptyset;$
> $\quad$ **WHILE** there exists j in $\{1, \ldots, n\} \setminus T$ such that either
> $\quad\quad\quad$ (a) $rank(\sigma_j) \leq rank(\sigma_{j'})$, for some $j' \in T$, or
> $\quad\quad\quad$ (b) $j \in \mathcal{F}$ and $\{i : \sigma_i \subseteq My_s \text{ for some } My_s \in \sigma_j\} \subseteq T$
> $\quad$ **DO**
> $\quad\quad\quad T \leftarrow T \cup \{j\}$
> $\quad$ **END WHILE.**

It can easily be verified that the set T produced by the above procedure has the form $\{1, \ldots, r_{k_1}\}$, for some $0 \le k_1 \le \ell$ and that it is an *admissible set of trapped places* according to Definition 2.3.

Notice that the number k_1 is an upper bound for the rank of any set σ_j, $j \in T$. Therefore, for each $j \in T$ we can put

$$\overline{\pi}_j =_{\text{Def}} \sigma_j.$$

To show that condition (C2) is satisfied, let y_s and π_j be trapped and such that $\overline{M}y_s \in \overline{\pi}_j$. Since $\overline{M}y_s = My_s$ and $\overline{\pi}_j = \sigma_j$, by (5), we have $F(s) = j$.

If $\pi_j(y_s) = 1$, i.e. $\sigma_j \subseteq My_s$, then $rank(\sigma_j) \le rank(My_s) < rank(\sigma_{F(s)})$. Thus $R(j) < R(F(s))$, showing that condition (C3) is also satisfied.

Next suppose that $y_s = \{y_t\}$ is in P. Then $My_s = \{My_t\}$, from which $\Pi(y_s) = \{F(t)\}$, proving condition (C4).

Let us put

$$FIN = \{j \in \{1, \ldots, n\} \ : \ \sigma_j \text{ is finite}\}.$$

It is straightforward to see that FIN is an admissable set of finite places, according to Definition 2.4.

Suppose now that $My_s \le My_t$. Then clearly

$$\max\{rank(\sigma_i) \ : \ i \in \Pi(y_s)\} \le \max\{rank(\sigma_j) \ : \ j \in \Pi(y_t)\}.$$

Therefore $s^* \le t^*$. Analogously, if $My_s < My_t$, then $s^* < t^*$. Moreover, if $i \in \mathcal{F}$, then the set σ_i has only elements of type My_s, $y_s \in V$. Hence $rank(\sigma_i) = rank(My_t) + 1$, where My_t is an element of maximum rank in σ_i. This implies that $R(i) = \max\{s^* : s \in V(\pi_i)\} + 1$, thus proving condition (C5).

Let $f = |FIN|$ and suppose that there are at most $p \in \{1, \ldots, n - f\}$ infinite sets σ_i of different cardinality. We can then define a function $\mathcal{C}$ from INF onto $\{0, \ldots, p - 1\}$ in such a way that $\mathcal{C}(j_1) < \mathcal{C}(j_2)$ if and only if $|\sigma_{j_1}| < |\sigma_{j_2}|$. Then clearly condition (C6) is true.

Finally, let SYS be the system defined in condition (C7) of Theorem 3.1. Then it can easily be verified that

$$\xi_j = |\sigma_j|, \quad \text{for } j \in FIN$$

is a positive integer solution of SYS, so that condition (C7) is satisfied too.

We have then showed that the conditions of Theorem 3.1 are necessary for P to be injectively satisfiable. In the next subsection we will prove the converse.

3.2 Sufficiency of conditions (C1)-(C7)

Again, let P be a normalized conjunction of MLSSRC with variables $V = \{y_1, \ldots, y_m\}$. Assume that there exist a set $\Pi = \{\pi_1, \ldots, \pi_n\}$ of places of P, a map $F : \{1, \ldots, m\} \to \{1, \ldots, n\}$, an admissible set of filled places $\mathcal{F}$, an increasing sequence of integers $0 = r_0 < r_1 < \ldots < r_\ell = n$, an admissible set of trapped places $T = \{1, \ldots, r_{k_1}\} \subseteq \Pi$, an assignment of sets $\overline{\pi} \subseteq V_{k_1}$ to the trapped places π, an admissible set of finite places FIN, and a map $C : INF \to \{0, 1, \ldots, n - f - 1\}$, such that conditions (C1)-(C7) are satisfied. In particular, let

$$\xi_i = \overline{\xi}_i , \quad i \in FIN$$

be a positive integer solution of the system SYS.

We will exhibit below a procedure that under such hypotheses constructs sets σ_i, $1 \leq i \leq n$, such that the assignment

$$M^{\bullet}y_s = \bigcup_{\pi_s(y_s)=1} \sigma_i \qquad (6)$$

is an injective model of P.

Let $k' \in \{k_1 + 1, \ldots, \ell + 1\}$ be the maximum integer such that $\{j : R(j) < k'\} \subseteq FIN$. Notice that if $k' = \ell + 1$ then $INF = \emptyset$. In this case or if $k' = k_1 + 1$ we put $k_2 = k' - 1$. Otherwise we introduce the function $S : \{1, \ldots, \ell\} \to \{0, 1\}$ defined by

$$S(h) = 1 \text{ if and only if } h = R(j), \text{ for some } j \in \mathcal{F} , \qquad (7)$$

for all $h = 1, \ldots, \ell$. So we define k_2 to be the integer in $\{k_1, \ldots, k' - 1\}$ such that $S(k_2 + 1) = 0$ and $S(h) = 1$ for all $k_2 + 1 < h \leq k'$. It will follow from the construction to be given below that $rank(\sigma_j) < \aleph_0$ if and only if $j \leq r_{k_2}$.

Let $\gamma > m$ be any finite ordinal such that

$$|\mathcal{V}_\gamma \setminus \mathcal{V}_{\gamma-1}| > (n + 1) \cdot \sum_{i \in FIN} \xi_i + m . \qquad (8)$$

Let

$$c = \max\{C(i) : i \in INF\}$$

and put:

$$\alpha(j) = \begin{cases} \gamma & \text{if } r_{k_1} < j \leq r_{k_2} \\ \aleph_c & \text{if } r_{k_2} < j \leq n . \end{cases} \qquad (9)$$

Let $B_{r_{k_1}+1}, \ldots, B_n$ be pairwise disjoint sets of rank γ and such that

$$|B_{r_{k_1}+1}| = \cdots = |B_n| = \sum_{i \in FIN} \overline{\xi}_i .$$

Notice that (8) guarantees that such sets can be found.

Let us also put

$$I_j = \{\alpha(j) + R(j)\} \cup B_j , \qquad (10)$$

for $j = r_{k_1} + 1, \ldots, n$. Observe that

$$|I_j| = \sum_{i \in FIN} \xi_i + 1 \qquad (11)$$

$$rank(I_j) = \alpha(j) + R(j) + 1. \qquad (12)$$

Moreover, we have the following lemma.

8

Lemma 3.1 *For all $j, j' \in \{r_{k_1}+1, \ldots, n\}, j \neq j'$,*

$$I_j \notin I_{j'}, I_j \neq I_{j'}$$

Proof. Suppose that $I_j \in I_{j'}$. Since $rank(I_j) = \alpha(j) + R(j) + 1$ it follows that $I_j \notin B_{j'}$. Thus $I_j = \alpha(j') + R(j')$, contradicting the fact that I_j is not an ordinal.

The second part of the assert follows from the disjointness of the sets B_j.

Let now

$$d_j = \begin{cases} \overline{\xi}_j - |\{s : F(s) = j\}| - 1 & \text{if } j \in FIN \\ \aleph_{C(j)} & \text{if } j \in INF \end{cases} \tag{13}$$

Then, for $j = r_{k_1}+1, \ldots, n$, there are sets A_j such that

$$A_j \subseteq \begin{cases} \mathcal{V}_\gamma \setminus \mathcal{V}_{m+1} & \text{if } j \in FIN \\ \mathcal{V}_{\aleph_c} \setminus \mathcal{V}_{\gamma+\ell+3} & \text{otherwise} \end{cases} \tag{14}$$

$$|A_j| = d_j \quad \text{if } d_j \geq 0 \tag{15}$$

$$A_j \cap \left(\bigcup_{i \neq j} A_i \cup \{I(i) : r_{k_1} < i \leq n\} \right) = \emptyset, \tag{16}$$

Following the increasing order of indices we put

$$\sigma_j = \begin{cases} \overline{\pi}_j & \text{if } j \in \mathcal{T} \\ \{M^\bullet y_s : F(s) = j\} & \text{if } j \in \mathcal{F} \\ \{M^\bullet y_s : F(s) = j\} \cup A_s \cup \{I_j\} & \text{otherwise} \end{cases} \tag{17}$$

where $M^\bullet y_s$ is inductively defined as in (6).

Notice that (17) is well given. Indeed, if $F(s) = j$, then, by (C4), $R(i) < R(j)$ for all $i \in \Pi(y_s)$. In particular, since R is nondecreasing, $i < j$, i.e. the set $M^\bullet y_s = \bigcup_{\pi_i(y_s)=1} \sigma_i$ is already defined when it is used in the definition of σ_j. Moreover, we have the following lemma.

Lemma 3.2 $I_j \neq M^\bullet y_s$ *for all* $r_{k_1} < j \leq n$ *and* $1 \leq s \leq m$.

Proof. It is enough to observe that if $\Pi(y_s) \cap INF \neq \emptyset$ then $|M^\bullet y_s| \geq \aleph_0$, whereas if $\Pi(y_s) \cap INF = \emptyset$ then $|M^\bullet y_s| < \sum_{i \in FIN} \xi_i$ On the other hand, by (11), $|I_j| = \sum_{i \in FIN} \xi_i + 1$. ∎

The sets σ_j's satisfy several useful properties. The most important are listed in the following lemma.

Lemma 3.3 *(A1)* $rank(\sigma_j) \leq m$, *for all* $1 \leq j \leq r_{k_1}$;

(A2) $rank(\sigma_j) = rank(\{I_j\}) = \alpha(j) + R(j) + 2$ *for all* $r_{k_1} < j \leq n$ *and* $rank(M^\bullet y_s) = \alpha(r_{s^\bullet}) + s^\bullet + 2$;

(A3) $\sigma_j \neq \emptyset$;

(A4) σ_j *is finite if and only if* $j \in FIN$ *and in this case* $|\sigma_j| = \overline{\xi}_j$;

(A5) $\sigma_j \cap \sigma_{j'} = \emptyset$, *for all* $j' \in \{1, \ldots, n\}$, $j' \neq j$;

9

(A6) $|\sigma_j| = \aleph_{C(j)}$ *if* $j \in INF$.

Proof. Properties (A1)-(A5) can be proved much in the same way as in Lemma 4.2 in [CC88]. The fact that $|\sigma_j| = \overline{\xi}_j$, for each $j \in FIN$, follows easily from (13), (14), (15), and (17). Property (A6) can be proved analogously. ∎

We are now ready to show that M^* is an injective model for P.

Injectivity of M^* and clauses of type (=).

From (C1), (A3) and (A5), it follows that M^* is injective. Moreover, by the definition of places, all clauses in P of type $x = y \cup z, x = y \setminus z$ are correctly modeled.

Clauses of type ({·}).

By (C4), if $y_s = \{y_t\}$ is in P, then $\Pi(y_s) = \{F(t)\}$. Thus by (17) $M^* y_s = \{M^* y_t\}$.

Clauses of type ($\leq, <$).

From (C5.1), if $y_s \leq y_t$ is in P, then $s^* \leq t^*$. Thus $M^* y_s \leq M^* y_t$ follows from (A2) and from the fact that the maps α and * (respectively defined in (9) and (4)) are nondecreasing. Analogously for clauses of type $y_s < y_t$.

Clauses of type ($|\ |_\leq$, $|\ |_<$).

Assume that $|y_s| \leq |y_t|$ occurs in P. Since $M^* y_s$ is finite if and only if $\Pi(y_s) \subseteq FIN$, the only two non trivial cases are :

(i) $\Pi(y_t) \subseteq FIN$. Then $\Pi(y_s) \subseteq FIN$. Therefore, from (A4), $|\sigma_j| = \overline{\xi}_j$, for all $j \in \Pi(y_s) \cup \Pi(y_t)$, so that $|M^* y_s| = \sum_{i \in \Pi(y_s)} \overline{\xi}_i$ and $|M^* y_t| = \sum_{j \in \Pi(y_t)} \overline{\xi}_j$. But by condition (C7.2), $\sum_{i \in \Pi(y_s)} \xi_i \leq \sum_{j \in \Pi(y_t)} \xi_j$, implying $|M^* y_s| \leq |M^* y_t|$.

(ii) $\Pi(y_s) \cap INF \neq \emptyset$. Then, by (C6), $\max\{C(j) : j \in \Pi(y_s) \cap INF\} \leq \max\{C(j) : j \in \Pi(y_t) \cap INF\}$. Thus, by (A6), $|M^* y_s| \leq |M^* y_t|$.

Analogously for clauses of type $|y_s| < |y_t|$.

This shows that M^* is an injective model for P, thus completing the proof of *Theorem* 3.1.

4   Final remarks

We conclude with the following remarks.

Remark 4.1 In the theory MLSSRC it is possible to express (1) natural numbers, (2) the *Finite* predicate, and (3) addition of cardinal numbers. Indeed,

(1) integers can be expressed in the following way:

$$0 = \emptyset \ , \ 1 = \{0\} \ , \ 2 = \{0, 1\} \ , \ 3 = \{0, 1, 2\} \ , \ \ldots$$

and in general

$$n + 1 = \{0, 1, \ldots, n\} \,,$$

where the finite enumeration operator $\{\cdot, \ldots, \cdot\}$ can easily be expressed in terms of repeated applications of the singleton and the binary union operators. So, literals like $x = n$, where n is a nonnegative constant, are expressible in MLSSRC.

(2) The predicate $Finite(x)$ [resp. $\neg Finite(x)$] is equivalent to the MLSSRC formula

$$x = \emptyset \vee (y \in x \wedge |x \setminus \{y\}| < |x|)$$

$$[\text{resp.} \ y \in x \wedge |x \setminus \{y\}| = |x| \ ].$$

(3) $|x| + |y| = |z|$ is equivalent to the conjunct:

$$|x'| = |x| \wedge |y'| = |y| \wedge x' \cap y' = \emptyset \wedge |z| = |x' \cup y'|,$$

where obviously $|v| = |w|$ is a short-hand for $|v| \leq |w| \wedge |w| \leq |v|$.

Remark 4.2 The theory MLSSRC remains decidable even if it is extended by the predicate $HF(s)$ which says that the set s is hereditarily finite. (We recall that a set is *hereditarily finite* if in any chain of type

$$s_r \in s_{r-1} \in \cdots \in s_1 \in s_0 = s \,,$$

s_r is finite, or, equivalently, $rank(s)$ is finite.)

In this case Theorem 3.1 has to be modified so as to include the following additional conditions:

Condition (C8). If $HF(y_s)$ occurs in P, then $\Pi(y_s) \subseteq FIN$.

Condition (C9). Let the map $S : \{1, \ldots, \ell\} \to \{0, 1\}$ be defined as in (7). Then there must exist an integer $k_2 \in \{k_1, \ldots \ell\}$ such that:

- if $HF(y_s)$ is in P and $i \in V(y_s)$, then $R(i) \leq k_2$;
- $j \in FIN$, for all j such that $R(j) \leq k_2$;
- $S(k_2 + 1) = 0$.

References

[Can88]   D. Cantone. Decision procedures for elementary sublanguages of set theory. X. Multilevel syllogistic extended by the singleton and powerset operators. *Journal of Symbolic Computation - Special issue*, to appear. New trends in Automated Mathematical Reasoning - Proceedings.

[CC88]   D. Cantone and V. Cutello. Decision procedures for elementary sublanguages of set theory. XV. Multilevel syllogistic extended by the predicate *finite* and the operators singleton and $pred_<$. *Rapporto di ricerca del Dipartimento di Matematica ed Informatica dell'Università di Udine n. 39*, 1988. Submitted to *J. Automated Reasoning*.

[CCF88]   D. Cantone, V. Cutello, and A. Ferro. Decision procedures for elementary sublan-
          guages of set theory. XIV. Three languages involving rank related constructs. In
          International Symposium on Symbolic and Algebraic Computation, to appear.

[CFMS87] D. Cantone, A. Ferro, B. Micale, and G. Sorace. Decision procedures for elementary
          sublanguages of set theory. IV. Formulae involving a rank operator or one occurrence
          of $\Sigma(x) = \{\{y\}|y \in x\}$. *Comm. Pure App. Math.*, XL:37–77, 1987.

[CS88]   D. Cantone and J.T. Schwartz. A set-theoretically based proof verifier. *Proposal to
          the National Science Foundation*, 1988.

[CS88]   D. Cantone and J.T. Schwartz. Decision procedures for elementary sublanguages of
          set theory. XI. Multilevel syllogistic extended by some elementary map constructs.
          Journal of Symbolic Computation - Special issue, to appear. New trends in Auto-
          mated Mathematical Reasoning - Proceedings.

[FOS80]   A. Ferro, E. Omodeo, and J.T. Schwartz. Decision procedures for elementary sub-
          languages of set theory. I. Multilevel syllogistic and some extensions. *Comm. Pure
          App. Math.*, XXXIII:599–608, 1980.

[GN72]   R.S. Garfinkel and G.L. Nemhauser. *Integer programming*. John Wiley & Sons, Inc.,
          New York, 1972.

[Jec78]   T.J. Jech. *Set theory*. Academic Press, New York, 1978.

[Sal75]   H.M. Salkin. *Integer programming*. Addison-Wesley Publishing Co., Inc., Reading,
          Mass., 1975.

NYU COMPSCI TR-440
Cantone, D
A decision procedure for
 set-theoretic formulae
 involving rank... c.1

NYU COMPSCI TR-440
Cantone, D
A decision procedure for
 set-theoretic formulae
 involving rank... c.1

DATE DUE	BORROWER'S NAME	

CPSIA information can be obtained
at www.ICGtesting.com
Printed in the USA
BVHW060322061118
532207BV00027B/5415/P

9 781333 235536